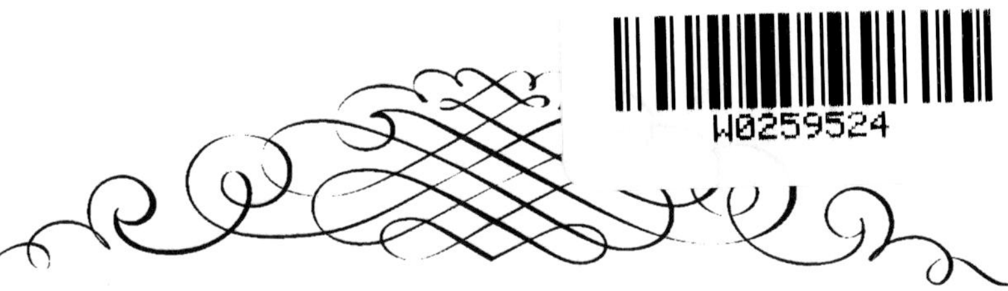

ISBN 978-1-332-26253-3
PIBN 10306003

Similar Books Are Available from
www.forgottenbooks.com

A History of Mathematics
by Florian Cajori

Differential Calculus for Beginners
by Joseph Edwards

The Calculus for Engineers
by John Perry

Integral Calculus for Beginners
With An Introduction to the Study of Differential Equations, by Joseph Edwards

On the Definition of the Sum of a Divergent Series, Vol. 1
by Louis Lazarus Silverman

A Treatise on Bessel Functions
And Their Applications to Physics, by Andrew Gray and G. B. Mathews

A Primer of Calculus
by Arthur Stafford Hathaway

Smithsonian Mathematical Formulae and Tables of Elliptic Functions
by Edwin P. Adams

A Treatise on Ordinary and Partial Differential Equations
by William Woolsey Johnson

Smithsonian Mathematical Tables
Hyperbolic Functions, by George F. Becker

The Mathematical Writings of Duncan Farquharson Gregory
by Duncan Farquharson Gregory

Mathematical Handbook
by Edwin P. Seaver

Elements of the Theory of Functions of a Complex Variable
With Especial Reference to the Methods of Riemann, by H. Durège

Vector Calculus
With Applications to Physics, by James Byrnie Shaw

Handbook of Engineering Mathematics
by Walter E. Wynne

Numerical Solution of the Boltzmann Equation
by Alexandre Joel Chorin

Engineering Applications of Higher Mathematics, Vol. 1
Problems on Machine Design, by Vladimir Karapetoff

Calculus
by Herman W. March

Vector Analysis
by Edwin Bidwell Wilson

Introduction to the Calculus of Variations
by William Elwood Byerly

A GROUP THEORETIC BRANCH AND BOUND ALGORITHM FOR THE ZERO-ONE INTEGER PROGRAMMING PROBLEM*

by

Jeremy F. Shapiro **

December 18, 1967

Working Paper 203-67

* This work was supported in part by Contract No. DA-31-124-ARO-D-209, U. S. Army Research Office (Durham)

** The author is indebted to Wayne W. Baxter for his invaluable computer programming contributions.

ABSTRACT

This paper contains a new algorithm for the zero-one integer programming problem. A given problem is solved first as a linear programming problem and an optimal basis is obtained. As shown by Gomory, the optimal basis is used to transform the integer problem into a new optimization problem over a finite abelian group. The group problem can be viewed as an unconstrained shortest route problem in a directed network and it is useful for two reasons: 1) the structure of the group is far simpler than that of the original problem, and 2) all feasible solutions to the original problem correspond to paths connecting a specified pair of nodes in the network. If one of the shortest route paths connecting the pair of nodes yields a feasible solution to the original problem, then the solution is optimal.

For a majority of zero-one problems, however, solving the unconstrained shortest route problem does not work. Constraints must then be added to the unconstrained problem making it very difficult to solve explicitly. The algorithm of this paper finds an optimal solution to the integer programming problem by implicitly enumerating all feasible solutions. Each partial solution in the implicit enumeration scheme is regarded as a path in the group network and tight lower bounds on the cost of an optimal completion are found by solving various residual shortest route problems. The group theoretic appraoch also gives insight into efficient branching procedures.

1. Introduction

In [12], Gomory showed how to transform the integer programming problem into a group optimization problem. The transformation requires that the integer programming problem be solved first as a linear programming (LP) problem. If B is an optimal basis for the LP problem, the group problem involves only the non-basic variables relative to B. This problem can be viewed as one in which we seek a minimal cost, non-negative integer, non-basic (optimal) correction[1] to the LP basic solution that makes it non-negative integer. The usefulness of the approach is due to the simple structure of the group problem when compared to that of the original integer programming problem. Moreover, the group setting appears to be a particularly good one for resolving the number theoretic difficulties of integer programming.

In particular, the group problem can be viewed ([25]) as a special shortest route problem with side constraints in a directed network with a finite number of nodes and arcs. The set of all feasible solutions to a given integer programming problem corresponds to a collection of paths connecting a specified node in the network to a second specified node. An unconstrained shortest route path connecting the specified pair of nodes is backtracked to yield a first attempt at an optimal correction to the LP optimal solution ([25]). If the

---

[1] A more precise definition of an optimal correction is given in section 2.

induced correction and the resulting LP basic variables constitute a feasible solution to the integer programming problem, then this solution is optimal. Sufficient conditions can be given ([12],[25]) on when an unconstrained shortest route path can be guaranteed to produce a feasible and thus optimal integer solution.

As discussed in [26], the class of problems for which the unconstrained shortest route solution will yield the optimal integer solution can be described qualitatively as steady-state. If b is the vector of constants in the integer programming problem, steady state means that the optimal LP solution $B^{-1}b$ is sufficiently large in each component to remain non-negative after the correction from the unconstrained shortest route or group problem is obtained.

If the unconstrained path fails to yield an optimal correction, there are several alternatives available. First, the group problem can be used to generate and identify good cuts for Gomory's cutting plane algorithm([13]). Another alternative is to find the kth shortest route in the network (k=2,3,4,...) connecting the specified nodes. In this case, convergence is through super-optimal infeasible solutions to a feasible and therefore optimal integer solution [27]. If the backtracked correction from the unconstrained shortest route path can almost be used to solve the original integer programming problem, then a third alternative discussed in [26] may be suggested. The backtracked correction "almost solves" the original problem if the resulting number of infeasibilities in the basic variables is small.

All of these extensions appear to be most useful for problems that are basically steady-state in character.

It is unfortunate that a very large and interesting class of integer programming problems are decidedly transient in nature; these are the zero-one problems. The most promising method to date for solving zero-one problems has been that of branch and bound, partly because the branch and bound approach is flexible enough to take advantage of the special structures of a given problem. Since an enormous variety of discrete optimization problems can be modeled as zero-one integer programming problems, it is to be expected that a general integer programming algorithm should often be inefficient when compared with a specially constructed, usually branch and bound algorithm. On the other hand, there may be value in the general formulation. This value would be derived from the insight into the number theoretic difficulties of any problem provided by the general formulation and approach. Viewed another way, one might say that branch and bound algorithms can easily incorporate and exploit macro-structures which are peculiar to a given problem, while the general algorithms attempt to analyze and resolve the number theoretic difficulties at the micro-level.

The purpose of this paper is the construction of an implicit enumeration algorithm for solving the constrained group optimization problem. Glover [10] has also discussed this problem. What appears novel here is the explicit use we make of the shortest route representation

of the group problem in developing new bounds. Moreover, an efficient sub-algorithm is constructed for solving the group problem with some but not all of the side constraints included. Finally, an explicit algorithm for the zero-one integer programming problem is given which fully integrates the group theoretic concepts and those of implicit enumeration.

Section 2 contains a brief review of the basic ideas on the transformation to the group problem from [12] and [25]. The following section begins with the statement of our problem in canonical form and the implicit enumerative scheme to be used in solving it. The concepts and methods of implicit enumeration are very clearly expounded by Professor Geoffrion in [6] and we shall make liberal use of his work. The remainder of section 3 is devoted to the development of explicit bounds from the shortest route group problem which are incorporated into the implicit enumeration scheme. The group problem is also used in the construction of the branching procedures.

Section 4 contains a concise statement of the algorithm and a numerical example of the algorithm is given in section 5. A partial list of computational results is presented in section 6. These results are for the algorithm of [22] for the unconstrained group problem. It is hoped that later versions of this paper will contain complete computational results.

Section 7 is devoted to final remarks and a discussion of areas for future research. Four sub-algorithms are given in the appendices.

2. Review and Terminology

The problem is written in initial canonical form as (cf [6])

$$\text{minimize } c'x'$$

subject to $$-b + A'x' \geq 0$$

(1) $$x'_j = 0 \text{ or } 1,$$

where c' is an n dimension vector of integers with component c'_j; b is an m dimensional vector of integers with component b_i; A' is an mxn dimensional vector of integers with generic column a'_j. The constraint relations are made into equations by adding surplus variables. This problem is then solved as the linear programming (LP) problem

$$\min \quad cx$$

(2) subject to $$Ax = b$$

$$x \geq 0$$

where $A = (A', -I)$ and $x = (x', s)$ and $s = (s_1, \ldots, s_m)$ is the vector of surplus variables. The generic column of A is denoted by a_j. Explicit upper bounds of the form $x'_j \leq 1$ are omitted in (2) because such constraints are quite often implicit in $Ax = b$. An alternative approach

is to use the upper bounding version of the simplex algorithm discussed in Chapter 18 of Dantzig [5].

We assume (2) has a bounded solution. In other words, assume that the simplex method applied to (2) yields a non-singular (optimal) mxm basis B with the properties

(i) $B^{-1}b > 0$

(ii) $c_j^* = c_j - c_B B^{-1} a_j \geq 0 \quad j=1,2,\ldots, m+n$

where c_B is the vector of cost coefficients corresponding to B. Without loss of generality, the columns of A are rearranged so that $B = (a_{n+1},\ldots,a_{n+m})$. Similarly, the non-basics are in $R = (a_1,\ldots,a_n)$.[1] Let (1) V_B denote the set of indices of the basic zero-one variables; (2) U_B denote the set of indices of the basic surplus variables; (3) V_R denote the set of non-basic zero-one variables; (4) U_R denote the set of surplus non-basic varibles.

Using the optimal basis B, transform the integer programming problem (1) into the equivalent form

$$\min \sum_{j=1}^{n} c_j^* x_j$$

subject to

$$y = b - \sum_{j=1}^{n} a_j x_j$$

(3)

$$x_j = 0 \text{ or } 1 \qquad \text{if } j \in V_R$$

$$x_j = 0,1,2,\ldots \qquad \text{if } j \in U_R$$

[1]Order the non-basics by decreasing value of c_j^*. This leads to computational savings in the analysis below (see [25]).

$$y_i = 0 \text{ or } 1 \qquad \text{if } i \in V_B$$

$$y_i = 0,1,2,\ldots. \qquad \text{if } i \in U_B$$

where $b = B^{-1}b$, and $a_j = B^{-1}a_j$. As discussed in [12] or [25], we transform this problem into a related group optimization problem. The group problem is[1]

$$\min \quad G_1(\alpha_b) = \sum_{j=1}^{n} c_j^* x_j$$

$$\text{subject to} \qquad \sum_{j=1}^{n} \alpha_j x_j = \alpha_b \pmod D$$

(4) x_j non-negative integer

where $D = |\det B|$

$$\alpha_j = D\{B^{-1}a_j - [B^{-1}a_j]\}, \ j=1,\ldots,n,$$

and

$$\alpha_b = D\{B^{-1}b - [B^{-1}b]\}$$

The group in question here is the finite abelian group G generated by the $\{\alpha_j\}_{j=1}^{n}$ with addition module D. Let f be the mapping from the lattice of integer vectors in m-space to the group; that is, the m dimensional vector of integers a is mapped into the group element f(a)

---

[1]We use bold face square brackets [ ] to denote "integer part of"; that is [a] is the largest vector of integers t such that $t < a$.

by the above transformation. The group has D elements and in general has a far simpler structure than the original integer programming problem for which it is derived. In fact, (4) resembles the knapsack problem. A further discussion of this point is deferred for the moment.

The importance of (4) to (3) is embodied in the lemma below. First a definition is required. A non-negative integer solution $(x_1^*,\ldots,x_n^*)$ of the non-basic variables is called an <u>optimal correction</u> (to the optimal LP solution to (2)) if $(x_1^*,\ldots,x_n^*,\ y_1^*,\ldots,y_n^*)$ is an optimal solution to (3) where

$$y^* = b - \sum_{j=1}^{n} a_j x_j^*$$

LEMMA 1: An optimal solution $(x_1^*,\ldots,x_n^*)$ to (4) is an optimal correction if

(5i) $$x_j^* \leq 1,\ j \in V_R$$

(5ii) $$y_i^* > 0,\ i=1,\ldots,m$$

(5iii) $$y_i^* < 1,\ i \in V_B.$$

<u>Proof</u>: Let $z^* = (z_1^*,\ z_2^*,\ldots,z_{m+n}^*) = (z_R^*,\ z_B^*)$ be an optimal solution to (3). Then z^* satisfies conditions (5i), (5ii), and (5iii) and moreover,

$$z_B^* = B^{-1}b - B^{-1}Rz_R^*$$

is a vector of integers. In other words, $B^{-1}b$ and $B^{-1}Rz_R^*$ differ by a vector of integers or

$$B^{-1}Rz_R^* \equiv B^{-1}b \pmod{1}.$$

The integer parts of the vectors $B^{-1}a_j$, $j=1,\ldots,n$, and $B^{-1}b$ can be dropped because they equal $0 \pmod 1$. Finally, we can clear fractions by multiplying both sides by $D = |\det B|$. The result is

$$\sum_{j=1}^{n} \alpha_j z_j \equiv \alpha_b \pmod{D}$$

where α_j and α_b are defined in (4). Thus z^* is a feasible solution to (4) which implies

(6) $$\sum_{j=1}^{n} c_j^* z_j^* \geq \sum_{j=1}^{n} c_j^* x_j^* \quad .$$

Moreover, the fact that x^* is a feasible solution to (4) and satisfies (5i), (5ii), and (5iii) implies it is feasible in (3), and thus optimal in (3) by (6).

Thus, if we solve (4) and the three conditions of lemma 1 hold, then we have solved (3) (or equivalently (1)). It is unlikely, however, that these conditions should hold for a given zero-one problem. One

obvious difficulty with (4) is that we allow x_j, $j\varepsilon V_R$, to take on integer values greater than one. Conversely, it is readily seen from the proof of lemma 1 that the problem we are most interested in solving is (4) with the proper side constraints; namely, conditions (5i), (5ii), and (5iii). Unfortunately, the addition of all these conditions to (4) makes it extremely difficult to solve explicitly. Our approach, therefore, will be to use branch and bound methods to solve (4) implicitly with conditions (5i), (5ii), (5iii) holding. Roughly speaking, the algorithm will implicitly test all solutions to (3) and will make use of (4) with a variety of weak side constraints in developing tight bounds and efficient branching procedures.

The usefulness of branch and bound methods becomes more apparent if (4) is viewed as a network optimization problem. To do this, we need to return to a fuller discussion of the group G generated by $\{\alpha_j\}_{j=1}^n$ in (4).

A useful isomorphic respresentation of G is

(7) $$G = Z_{q_1} \oplus Z_{q_2} \oplus \cdots \oplus Z_{q_r}$$

where

$$Z_{q_i} = \text{residue class of the integers modulo } q_i,$$

and

$$q_i | q_{i+1}, \; i=1,\ldots,r-1 \text{ and } \prod_{i=1}^{r} q_i = D$$

Moreover, given a finite abelian group, there is a unique representation of the form (7). Thus, G can be represented as a collection of D r-tuples $(\ell_1,\ldots,\ell_r)$ with addition of the r-tuples mod $(q_1,\ldots,q_r)$. The usefulness of this representation is two-fold: (1) in general, r is considerably less than m and often equals one; (2) this representation of G exposes the basic structure of the group, a structure which can be exploited.

A computational algorithm for achieving the representation (7) is based on a matrix diagonalization procedure. The theory behind this algorithm is described in [24]. The algorithm has already been programmed for the IBM 360-ASP System at M.I.T. and representative results are discussed in section 6.

The r-tuples $\{(\ell_1,\ldots,\ell_r);\ \ell_1 = 0,\ldots,\ q_1-1,\ldots,\ \ell_r = 0, 1,\ldots,\ q_r-1\}$ are lexicographically ordered by the rule $(\ell_1,\ldots,\ell_r) \prec (t_1,\ldots,t_r)$ if $\ell_s < t_s$ where s is the smallest index i such that $\ell_i < t_i$. Let λ_k be a typical r-tuple and $G = \{\lambda_k\}_{k=0}^{D-1}$ where $\lambda_o = \theta$. Strictly speaking, the m vectors α_j and α_b computed in (4) should be distinguished from their isomorphic representation as r-tuples. We will not make this distinction because only the representation (7) will be used below.

It is convenient to view (4) as the problem of finding a shortest route path connecting a pair of nodes in a related network. If (4) fails to solve (3) (see lemma 1), the optimal correction we seek still corresponds to a path in the network connecting the same pair of nodes. In [27], White has developed an algorithm for finding this path explicitly.

Our approach here can be viewed as implicit in the sense of Balas' algorithm ([1], [7]).

The network of problem (4) which we call the group network, is $\Gamma = [M,R]$. The set M contains D nodes, one for each element λ_k of G. The arcs in R are of the form $(\lambda_k - \alpha_j, \lambda_k)$, k=1,...,D-1. An arc cost c_j^* is associated with each arc $(\lambda_k - \alpha_j, \lambda_k)$. Finding an optimal solution to (4) is equivalent to finding a shortest route path from θ to α_b in μ. If $\mu 1(\alpha b)$ is a shortest route path from θ to α_b, then $x^1 = (x_1^1,\ldots,x_n^1)$ is an optimal solution to (4) where x_j^1 = number of times an arc of the form $(\lambda_k - \alpha_j, \lambda_k)$ appears in $\mu^1(\alpha_b)$. It is important to note that we have a special shortest route problem because the same type of arc is drawn to each node. A specially constructed algorithm which exploits this structure is given in [25] and is repeated in Appendix A. In the course of finding $G_1(\alpha_b)$, the algorithm also finds $G_1(\lambda_k)$, k=0,1,...,D-1, the value of a shortest route path from θ to λ_k in Γ. The corresponding paths will be denoted by $\mu_1(\lambda_k)$ and the backtracked solutions by $x^1(\lambda_k)$. We remark that an almost identical shortest route network representation of the knapsack problem was developed in [23].

In order to facilitate the analysis below, augment Γ by a single node $\lambda_D = (q_1, q_2,\ldots,q_r)$ and by arcs $(\lambda_D - \alpha_j, \lambda_D)$, j=1,...,n, with arc costs c_j^*. The shortest route path from θ to λ_D is the least cost circuit in the network. This path is also found by the algorithm of [25] as stated in Appendix A.

---

[1]If $\alpha_j=\theta$, then the arc is (θ,λ_D) with arc cost c_j^*.

We summarize the discussion up to this point in terms of a corollary to lemma 1,

Corollary 1: Any feasible correction z_R to (3) corresponds to a path in Γ connecting θ to α_b. Moreover, (4) with the side constraints (5i), (5ii) (5iii) is equivalent to (3) in the sense that a feasible solution to (4) with these side constraints is feasible in (3) and vice versa.

Proof: The proof is immediate from the proof of lemma 1.

3. Bounding and Branching Procedures

We have seen that the original integer programming problem (3) in the group theoretic setting becomes the problem of finding a shortest route path in Γ connecting θ to α_b that satisfies conditions (5i), (5ii), and (5iii) of Lemma 1. The y_i^* in lemma 1 are uniquely determined from the x_j^* by the constraint equations of (3). We will construct an algorithm for finding such a path by implicitly enumerating all corrections.

In order to facilitate the branch and bound approach, transform the surplus variables x_j, $j \varepsilon U_R$, to zero-one variables. Let u_j be an upper bound[1] on x_j in (1) and $v_j = [\log_2 u_j]$. Define the zero-one variables δ_{jk} by

(8) $$x_j = \sum_{k=0}^{v_j} 2^k \delta_{jk}$$

To facilitate the analysis below, for each non-basic surplus variable x_j, $j \varepsilon U_R$, let

$$a_{j_k} = 2^k \bar{a}_j, \quad k=0,1,2,\ldots, \quad v_j,$$

$$c^*_{j_k} = 2^k c^*_j, \; k = 0,1,2,\ldots,v_j$$

---

[1] It is easily seen from problem (1) that such u_j exist.

and

$$\alpha_{j_k} \qquad f(a_{jk}) = 2^k \alpha_j$$

Problem (3) becomes

$$\min \quad \sum_{j\varepsilon V_R} c_j^* x_j \qquad \sum_{j\varepsilon U_R} \sum_{k=0}^{v_j} c^*_{j_k} \delta_{jk}$$

(9) subject to $$y = b - \sum_{j\varepsilon V_R} a_j x_j \qquad \sum_{j\varepsilon U_R} \sum_{k=0}^{v_j} a_{j_k} \delta_{jk}$$

$$x_j = 0 \text{ or } 1, \quad j\varepsilon V_R; \quad \delta_{jk} = 0 \text{ or } 1, \quad j\varepsilon U_R \;;$$

$$y_i = 0 \text{ or } 1, \quad i\varepsilon V_B; \quad y_i = 0,1,2,\ldots, \quad i\varepsilon U_B$$

It is important at this point to develop and review the systematic implicit enumeration scheme described in [7]. A partial solution S to (3) is an assignment of feasible values to a subset of the non-basic variables $\{x_1,\ldots,x_n\}$. The variables not included in the subset of the partial solution are called free variables. The current best solution at any point during computation is denoted by $\hat{x}$ with value $\hat{z} = c\hat{x}$ and is called the <u>incumbent</u>.

Given the partial solution S, the algorithm attempts either to continue S to an optimal completion or to ascertain that S has no feasible completion better than x. If either of these contingencies occurs, then S is said to be <u>fathomed</u>; that is, all feasible completions of S have been

implicitly enumerated. In this case we climb up the tree of enumerated solutions and free some of the fixed variables in S. Otherwise, a free variable relative to S is fixed and we descend the tree of enumerated solutions.

A non-redundant and implicitly exhaustive sequence of partial solutions is generated if the following rules are used. As discussed by Geoffrion, let j (j_k) denote $x_j = 1$ ($\delta_{jk} = 1$) and j ($-j_k$) denote j=0 ($\delta_{jk} = 0$).[1] A partial solution S is described by a series of such symbols.

The procedure for implicit enumeration of all corrections in (9) is[2]

Step 0: Put S = ϕ.

Step 1: Attempt to fathom S. Is the attempt successful?

YES: If the best feasible completion of S has been found and it is better than the incumbent solution, store it as the new incumbent. Go to Step 3.

NO: Go to Step 2:

Step 2: Augment S on the right by j(j_k) or -j($-j_k$) where x_j(δ_{jk}) is any free variable. Return to Step 1.

Step 3: Locate the rightmost element of S which is not underlined. If none exists, terminate; otherwise, replace the element by its underlined complement and delete all elements to the right. Return to Step 1."

---

[1] To avoid needless repetition, henceforth we shall use only j to denote a generic index in (9).

[2] See Geoffrion [7; figure 1].

Let S be a partial solution to (9). If S is feasible, then the best completion of S is to take $x_j = 0$, $j \notin S$, since $c_j^* \geq 0$. Denote this completion of S by x^S. Suppose S is not feasible. In this case we can develop a series of tests using both (9) and (4) for fathoming S. Define

$$T^S = \{j \text{ free: } c^*x^S + c_j^* < z \text{ and either}$$

$$(1)\quad a_{ij} > 0 \text{ for some } i \text{ such that } y_i^S < 0 \text{ or}$$

$$(2)\quad a_{ij} < 0 \text{ for some } i \text{ such that } y_i^S > 1 \text{ and } i \varepsilon V_B \},$$

$$\text{where}\quad y^S = b - \sum_{j \varepsilon V_R} a_j x_j^S - \sum_{j \varepsilon U_R} \sum_{k=0}^{v_j} \bar{a}_{j_k} \delta_{jk}^S$$

If T^S is empty, it is clear that no feasible completion of S is better than the incumbent and S is fathomed.

Similarly, there is no feasible completion of S better than the incumbent either if

$$(11)\quad y_i^S + \sum_{\substack{j \varepsilon T^S \\ \text{and} \\ j \varepsilon V_R}} \max\{0, a_{ij} + \sum_{\substack{j_k \varepsilon T^S \\ \text{and} \\ j_k \varepsilon U_R}} \max\{0, a_{ij_k} < 0$$

for some $y_i^S < 0$; or if

$$(12)\quad y_i^S + \sum_{\substack{j \varepsilon T^S \\ \text{and} \\ j \varepsilon V_R}} \min\{0, a_{ij} + \sum_{\substack{j_k \varepsilon T^S \\ \text{and} \\ j_k \varepsilon U_R}} \min\{0, a_{ij_k} > 1$$

for some $y_i^S > 1$ and $i \epsilon V_B$.

If these two tests do not result in S being fathomed, then we turn to the group problem (4) for further fathoming tests. Let α_S be the group element corresponding to S; that is, if

$$j_1, j_2, \ldots, j_v, \quad j_{k_1}, j_{k_2}, \ldots, j_{k_u}$$

are variables at a level of one in S, then

$$\alpha_S = \sum_{i=1}^{v} \alpha_{j_i} + \sum_{i=1}^{u} \alpha_{j_{k_i}} \tag{13}$$

It is clear by our discussion in section 2 that the optimal completion of S can be found by finding the least cost path in Γ connecting α_S to α_b such that

(i) only free (non-basic) variables relative to S are used as arcs,

(ii) the free (non-basic) bivalent variables are used at most once,

(iii) the resulting basic variable values from the completed solution are non-negative; that is $y_i > 0$, $i=1,\ldots,m$,

(iv) $y_i \leq 1$, $i \epsilon V_B$.

As before, an explicit solution to this shortest route problem with side constraints is difficult to find. Our approach here is to compute super-optimal paths by ignoring some or all of the side constraints. The hope is either that one of these paths will provide a lower bound on the cost of an optimal completion which exceeds the cost of the incumbent, or

that one will provide a feasible and therefore optimal completion.

In particular, there are three different lower bounds on the cost of an optimal completion of S. The first is $G_1(\alpha_b-\alpha_S)$ which is the cost of an optimal solution to (4) with right hand side $\alpha_b - \alpha_S$. The cost $G1(\alpha_b - \alpha_S)$ and the optimal path $\mu 1(\alpha_b - \alpha_S)$ are computed when IP Algorithm I is applied to (4). Note that it is possible that $\alpha_S = \alpha_b$ but S is not a feasible connection to (9). In this case, we must continue S by a (least cost) circuit from node α_b back to node α_b. This was the reason for the addition of node D to Γ in the previous section. In order to avoid excessive repetition below, we state now that for i=1,2,3, the bounds are denoted by $G_i(\alpha_b - \alpha_S)$, the backtracked paths in Γ are denoted by $\mu_i(\alpha_b - \alpha_S)$, and the corresponding corrections are denoted by $x^i(\alpha_b - \alpha_S)$

It is evident that S is fathomed if

$$(14) \qquad c^*x^S + G1(\alpha_b - \alpha_S) > \hat{z}$$

since the sum on the left is a lower bound on the value of an optimal completion of S. If the inequality (14) does not hold, but $x^S + x^1(\alpha_b - \alpha_S)$ is a feasible correction in (9), then S is fathomed and the new incumbent $\hat{x} = x^S + x^1(\alpha_b - \alpha_S)$ with value

$$z = c^*x^S + G1(\alpha_b - \alpha_S).$$

If knowledge of $G1(\alpha_b - \alpha_S)$ and $\mu_1(\alpha_b - \alpha_S)$ does not lead to a

fathoming of S, we proceed to the computation of the higher lower bound $G_2(\alpha_b - \alpha_S)$ on the cost of an optimal completion to S. $G_2(\alpha_b - \alpha_S)$ is the cost of a least cost path from α_S to α_b using only the free variables relative to S. More explicitly,

$$G_2(\alpha_b - \alpha_S) = \min \sum_{\substack{j=1 \\ \text{and} \\ j \notin S \cdot V_R}}^{n} c_j^* x_j$$

subject to

$$(15) \qquad \sum_{\substack{j=1 \\ \text{and} \\ j \notin S \cdot V_R}}^{n} \alpha_j x_j \quad \alpha_b - \alpha_S \ (\text{mod } D)$$

$$x_j \text{ non-negative integer.}$$

It is important to note that we include all non-basic surplus variables as free variables relative to S. In so doing, we ignore the upper bounds on the surplus variables. If these bounds are exceeded, the resulting solution will be infeasible and thus there is no loss of generality. As before, if

(16) $\quad c^*x^S + G_2(\alpha_b - \alpha_S) \geq z,$

or

(17) $\quad c^*x^S + G_2(\alpha_b - \alpha_S) < z$ and $x^S + x^2(\alpha_b - \alpha_S)$ is a feasible correction

then S is fathomed. In the latter case, set $z = c^*x^S + G_2(\alpha_b - \alpha_S)$ and $\hat{x} = x^S + x^2(\alpha_b - \alpha_S)$.

The reader will have noticed that the bounds $G_1(\alpha_b - \alpha_S)$ and $G_2(\alpha_b - \alpha_S)$

were computed without the explicit constraints that $x_j < 1$, $j \epsilon V_R$. Thus a tighter lower bound $G_3(\alpha_b - \alpha_S) \geq G_2(\alpha_b - \alpha_S)$ on the cost of an optimal completion of S is

$$G_3(\alpha_b \quad \alpha_S) \quad \min \sum_{\substack{j=1 \\ \text{and} \\ j \notin S \cup V_R}}^{n} c_j^* x_j$$

subject to

$$\sum_{\substack{j=1 \\ \text{and} \\ j \notin S \cup V_R}}^{n} \alpha_j x_j \quad \alpha_b \quad \alpha_S \pmod{D}$$

(18)

$$x_j = 0 \text{ or } 1, \quad j \epsilon V_R \cap S^c; \qquad x_j \text{ non-negative integer}, \ j \epsilon U_R.$$

Problem (18) cannot be solved by the algorithm of [25]. A new algorithm is given in Appendix B for solving this shortest route (group) problem when some of the arcs can be used at most once.

If the set of free variables is sufficiently small, it may be that $\alpha_b - \alpha_S$ cannot be spanned by the free variables, particularly when the bivalent free variables are used at most once. A sufficient condition for this to occur is that the free variables lie within or generate a proper subgroup of G that does not contain $\alpha_b - \alpha_S$. For example, suppose the canonical representation (7) of G consists of more than one cyclic subgroup. If the free variables can be represented by some partial direct sum of

the Z_{q_i}, but $\alpha_b - \alpha_S$ cannot, then S cannot be feasibly completed. As a numerical illustration, suppose $\alpha_b - \alpha_S = (1,2,7) \bmod (2,4,12)$ and the group identities of the free variables relative to S are (0,3,5), (0,1,3), and (0,1,4).[1]

There is an additional group theoretic lower bound on the cost of a completed partial solution S which is sometimes available. This lower bound will not be used here and it is presented primarily because it is related to the algorithm in [26]. The first step is to net out x^S from (9) with the result that (9) becomes

$$\min \sum_{j \in V_R \cap S^c} c_j^* x_j + \sum_{j_k \in U_R \cap S^c} c_{j_k}^* \delta_{jk}$$

subject to

$$y \quad b' \quad \sum_{j \in V_R \cap S^c} a_j x_j \quad \sum_{j_k \in U_R \cap S^c} a_{j_k} \delta_{jk}$$

(19) $$x_j = 0 \text{ or } 1; \quad \delta_{jk} = 0 \text{ or } 1;$$

$$y_i = 0 \text{ or } 1,\ i \in V_B: \quad y_i = 0,1,2,\ldots,\ i \in U_B,$$

where

$$\overline{b}' = b - \sum_{j \in V_R \cap S} a_j x_j \quad \sum_{j_k \in U_R \cap S} a_{j_k} \delta_{jk}$$

The new bound can be found whenever there is at least on $b_i' < 0$. In this case, the LP solution $y_i = b_i'$, $i=1,\ldots,m$, to (19) is not feasible.

[1] If the group G has a multi-dimensional canonical representation, the above shortest route problems could be solved for any group formed by taking a partial direct sum of the Z_{q_i}. The resulting shortest route values would be lower bounds on the shortest route values for the full group. This would be particularly useful when D is overly large.

Reoptimize LP(19) using the dual simplex algorithm since $c_j^* > 0$, $j=1,\ldots,n$. If the value of the new optimal LP solution plus c^*x^S is greater than z, S is fathomed. Otherwise, repeat the procedure outlined in section 2 using this new LP basis to convert (19) to a new group problem. If $\tilde{G}(\alpha_b - \alpha_S)$ (the tilde is used to denote the new group) is the solution to the new problem (4), then $c^*x^S + \tilde{G}(\tilde{\alpha}_b - \tilde{\alpha}_S)$ is a lower bound on the cost of an optimal completion of S. One cannot, of course, make a general comparison between the cost $G_2(\alpha_b - \alpha_S)$ and $G(\alpha_b - \tilde{\alpha}_S)$. Another new bound can be found by solving (18) for the new group.

As a final remark about lower bounds, it is important to note that the bound $G_2(\alpha_b - \alpha_S)$ (and in some cases the bound $G_3(\alpha_b - \alpha_S)$) need not be recomputed for each partial solution. In particular, suppose $S^{k+1},\ldots,S^{k+t}$ are all continuations of the partial solution S^k, and the algorithm in Appendix A was used to solve (15) for $S=S^k$. Letting the bounds be similarly superscripted, it is easy to see that

$$(20) \qquad G_2^{k+t}(\alpha_b-\alpha_{S^{k+t}}) \geq G_2^k(\alpha_b-\alpha_{S^{k+t}}) ,$$

since the free variables relative to S^{k+t} are a subset of the free variables relative to S^k. Recall that the algorithm finds $G_2^k(\lambda_\ell)$, $\ell=0,1,\ldots,D$, in the course of finding $G_2^k(\alpha_b-\alpha_{S^k})$, and hence $G_2^k(\alpha_b-\alpha_{S^{k+t}})$ is known. Thus if

$$c^*x^{S^{k+t}} + G_2^k(\alpha_b-\alpha_{S^{k+t}}) \geq z,$$

S^{k+t} is fathomed. Of course, if the path yielding $G_2^k(\alpha_b - \alpha_{S^{k+t}})$ involves only non-basic variables which are free relative to S^{k+t}, then

$$G_2^{k+t}(\alpha_b - \alpha_{S^{k+t}}) = G_2^k(\alpha_b - \alpha_{S^{k+t}}).$$

As for $G_3^{k+t}(\alpha_b - \alpha_{S^{k+t}})$, if the algorithm of Appendix B finds $G_3^k(\alpha_b - \alpha_{S^{k+t}})$ in the course of finding $G_3^k(\alpha_b - \alpha_{S^k})$, then the same inequality holds. This will generally be the case only when

$$G_3^k(\alpha_b - \alpha_{S^{k+t}}) = G_2^k(\alpha_b - \alpha_{S^{k+t}})$$

and the new information is relatively weak.

It is a simple matter to use the implicit enumeration scheme to keep track of when an intermediate bound such as $G_2^k(\alpha_b - \alpha_{S^{k+t}})$ is available. Whenever $G_2(\alpha_b - \alpha_S)$ is calculated for some partial solution S, overline the rightmost element of S which has one of the four forms, j, $-j$, $\underline{j}$, $-\underline{j}$. Remove all overlines to the left and erase the corresponding stored results.[1] Store $G_{2,j}(\lambda_k)$, and $x^{2,j}(\lambda_k)$, $k=0,1,\ldots,D$ which result from the application of the algorithm of Appendix A to (15).

Let S' be any other partial solution. If there is an element in S' which is overlined, then as described above, attempt to use the stored results corresponding to this element to fathom S'. Finally, whenever

---

[1]These results could be saved. We have erased them for ease of exposition and also because of possible storage limitations.

a partial solution is fathomed and elements from it are deleted, erase any stored results corresponding to deleted elements.

Although the above procedures in the group setting can yield optimal completions of a partial solution S, their primary usefulness is in providing tight lower bounds on the cost of such a completion. For this reason, it appears uneconomical to calculate the bounds from the group problem before an initial feasible solution to (9) is found. If an initial feasible solution is not provided, there are two methods for obtaining one. The first is to use the algorithm in [7] on problem (9) in the form (1) until an initial feasible solution is found. A second alternative is to use a non-optimizing backtracking algorithm to find paths from θ to α_b in the group. Each path selected is tested for feasibility. The algorithm and a short discussion of its properties are given in Appendix D.

If all of the fathoming tests discussed above fail, then a partial solution S needs to be continued by setting some free variables x_j or δ_{jk} to 0 or 1. If S seems to be a "good" partial solution in the sense that there is a relatively high expectancy that S will lead to an improvement over the incumbent, then the free variable should be chosen accordingly. Conversely, if it seems very unlikely that S will lead to an improvement over the incumbent, then a different rule is suggested. The problem of basing these heuristic ideas on a formal and rigorous model is beyond the scope of the present paper.

The approach here will be the following. Let

$$R^S = \{j: \; j \epsilon T^S \text{ and } c^*x^S + c_j^* + G_2(\alpha_b - \alpha_S - \alpha_j) < \hat{z}\ \} .$$ [1]

If R^S is empty, then S is fathomed. If R^S contains all the free variables, then we choose to set to one some free variable which minimizes the sum of the infeasibilities; viz.,

$$(21) \quad \min_{j \epsilon S^c} \; \{ \sum_{i=1}^{m} \max \{-(y_i^S + a_{ij}),0\} + \sum_{\substack{i=1 \\ \text{and} \\ i \epsilon V_B}}^{m} \max \{(y_i^S + a_{ij} - 1), 0\}\}$$

In case of ties, choose any free variable with minimal c_j^*. If R^S is not empty and does not contain all the free variables, then compare

$$\delta \; = \hat{z} - c^*x^S - G_2(\alpha_b - \alpha_S) > 0$$

with a prespecified parameter δ_0. If $\delta > \delta_0$, choose x_j=1 for j which satisfies (21); ties are broken as before. If $\delta < \delta_0$, select a free variable x_j for which

$$(22) \qquad c^*x^S + c_j^* + G_2(\alpha_b - \alpha_S - \alpha_j) > z$$

and set $x_j = 0$. Moreover, underline the rightmost element $-\underline{j}$ since (22) indicates that the continued solution {S,j} is automatically fathomed. In particular, choose that free variable x_j such that (22) holds according to the criterion (21). Ties in this case may be broken arbitrarily.

[1] If $\alpha_b-\alpha_S-\alpha_j=\theta$, and {S,j} is feasible in (9), then consider $G_2(\alpha_b-\alpha_S-\alpha_j)=0$. If {S,j} is not feasible in (9), consider $G_2(\alpha_b-\alpha_S-\alpha_j) = G_2(\lambda_D)$.

The criterion (21) reflects the prior feeling that the corrections from the group problem to be found for the next partial solution S' are just as likely to increase as decrease each $y_i^{S'}$. A more sophisticated criterion would be to scan the a_{ij}, j free, and then attempt to center accordingly the $y_i^{S'}$.

There are three additional comments to be made about the set R^S. First, for $S=\phi$, the variables not in R^S can be dropped from the problem. Similarly, if $S=S^k$ is a continuation of S^{k-1}, then $R^S=R^{S^k}$ can be replaced by $R^{S^{k-1}} \cap R^{S^k}$. Finally, we can replace T^S by $R^S \cap T^S$ in (11) and (12).

This completes our discussion of the bounding and branching procedures for the algorithm of this paper. In the next section we amalgamate all of the above components into an algorithm.

4. Group Theoretic Branch and Bound Algorithm for the Zero-One Integer Programming Problem

The algorithm below begins after the following initial steps are performed. First, (2) is solved and the optimal LP basis B is used to convert (2) to (3). Problem (3) is then transformed to the group problem in canonical form and IP Algorithm I is used to find $G1(\alpha_b)$. Section 6 contains representative times for these computations. If $x^1(\alpha_b)$ (or any backtracked solution corresponding to any path from θ to α_b with value $G1(\alpha_b)$) is a feasible correction in (3), then it is optimal and we are done. Otherwise, the results of IP Algorithm I are stored, problem (3) is converted to the form of (9), and the algorithm below is initiated.

Step 0 (Initialization) If a finite upper bound on the value of an optimal solution to (9) is known, set $\hat{z}$ to this upper bound. If such a bound is not known, use the Backtracking Algorithm (see Appendix D) to find an initial feasible solution and initial upper bound. Set $S=\phi$ and

$$R^{\phi} = \{j : c_j^* + G1(\alpha_b - \alpha_S) < \hat{z}\}$$

Drop from the problem any variable not in R^{ϕ}. Go to (1.1) with $R=R^{\phi}$.

Step 1 (Bounding Procedures)

(1.1) If x^S is a feasible correction in (9), S is fathomed; if $c^*x^S < z$, $\hat{z} \leftarrow c^*x^S$ $\hat{x} \leftarrow x^S$ and go to (2.2). Otherwise, go to (1.2).

(1.2) Define

$T^S = \{j$ free: $j \in R$ and either (1) $a_{ij} > 0$ for some i such that $y_i^S<0$, or (2) $a_{ij}<0$ for some i such that $y_i^S>1$ and $i \in V_B$, where $y^S = \bar{b} - \sum_{j=1}^{n} \bar{a}_j x_j^S$.

If $T^S = \phi$, S is fathomed; go to (2.2). Otherwise, if either

$$y_i^S + \sum_{\substack{j \in T^S \\ \text{and} \\ j \in V_R}} \max\{0, a_{ij}\} + \sum_{\substack{j_k \in T^S \\ \text{and} \\ j_k \in U_R}} \max\{0, a_{ij_k}\} < 0$$

for some $y_i^S<0$; or if

$$y_i^S + \sum_{\substack{j \epsilon T^S \\ \text{and} \\ j \epsilon V_R}} \min\{0, a_{ij}\} + \sum_{\substack{j_k \epsilon T^S \\ \text{and} \\ j_k \epsilon U_R}} \min\{0, a_{ij_k}\} > 1$$

for some $y_i^S > 1$ and $i \epsilon V_B$, then S is fathomed. Go to (2.2). If all of the above tests fail, go to (1.3). Retrieve $_{G1}(\alpha_b \quad \alpha_S)$. If

$$c^* x^S + {}_{G1}(\alpha_b - \alpha_S) \geq z,$$

S is fathomed; go to (2.2). if

$$c^* x^S + G_1(\alpha_b - \alpha_S) < \hat{z}$$

and $x^S + x^1(\alpha_b - \alpha_S)$ is a feasible correction in (9), S is fathomed;

$$z \leftarrow c^* x^S + {}_{G1}(\alpha_b - \alpha_S), \quad \hat{x} \leftarrow x^S + x^1(\alpha_b - \alpha_S) \quad ,$$

and go to (2.2). Otherwise, check $x^1(\alpha_b - \alpha_S)$ to see if $_{G1}(\alpha_b - \alpha_S) = G_2(\alpha_b - \alpha_S)$ and $_{G1}(\alpha_b - \alpha_S) = G_3(\alpha_b - \alpha_S)$. If neither equality holds, go to (1.4). If only the former equality holds go to (1.6). If both equalities hold, go to (2.1) with $\delta = z - c^* x^S - {}_{G1}(\alpha_b - \alpha_S)$.

(1.4) If there is no overlined element in S, go directly to (1.5). If there is an overlined element j, retrieve $G_{2,j}(\alpha_b - \alpha_S)$. If

$$c^*x^S + G_{2,j}(\alpha_b-\alpha_S) > z,$$

S is fathomed; go to (2.2) If

$$c^*x^S + G_{2,j}(\alpha_b-\alpha_S) < z,$$

and $x^S + x^{2,j}(\alpha_b-\alpha_S)$ is a feasible correction in (9), S is fathomed; $z \leftarrow c^*x^S + G_{2,j}(\alpha_b-\alpha_S)$, $x \leftarrow x^S + x^{2,j}(\alpha_b-\alpha_S)$, and go to (2.2). Otherwise, check $x^{2,j}(\alpha_b-\alpha_S)$ to see if $G_{2,j}(\alpha_b-\alpha_S) = G_2(\alpha_b-\alpha_S)$ and $G_{2,j}(\alpha_b-\alpha_S) = G_3(\alpha_b-\alpha_S)$. If neither equality holds, go to (1.5). If only the former equality holds, go to (1.6). If both equalities hold, go to (2.1) with $\delta = z - c^*x^S - G_{2,j}(\alpha_b-\alpha_S)$.

(1.5) Use IP Algorithm I to compute $G_2(\alpha_b-\alpha_S)$. Store the results and overline the right most element of S. If

$$c^*x^S + G_2(\alpha_b-\alpha_S) > \hat{z},$$

S is fathomed; go to (2.2). If

$$c^*x^S + G_2(\alpha_b-\alpha_S) < z$$

and $x^S + x^2(\alpha_b-\alpha_S)$ is a feasible correction in (9), S is fathomed; $z \leftarrow c^*x^S + G_2(\alpha_b-\alpha_S)$ and $\hat{x} \leftarrow x^S + x^2(\alpha_b-\alpha_S)$. Otherwise, check $x^2(\alpha_b-\alpha_S)$ to see if $G_2(\alpha_b-\alpha_S) = G_3(\alpha_b-\alpha_S)$. If not, go to (1.6). If the equality

holds, go to (2.1) with $\delta = \hat{z} - c^*x^S - G_2(\alpha_b-\alpha_S)$

(1.6) Use the algorithm in Appendix B to compute $G_3(\alpha_b-\alpha_S)$ If

$$c^*x^S + G_3(\alpha_b-\alpha_S) \geq z,$$

S is fathomed; go to (2.2). If

$$c^*x^S + G_3(\alpha_b-\alpha_S) < z$$

and $x^S + x^3(\alpha_b-\alpha_S)$ is feasible in (9), S is fathomed; z $c^*x^S + G_3(\alpha_b-\alpha_S)$ $\hat{x} \leftarrow x^S + x^3(\alpha_b-\alpha_S)$; and go to (2.2). Otherwise, go to (2.1) with $\delta = z - c^*x^S - G_3(\alpha_b-\alpha_S)$.

Step 2 (Branching Procedures)

(2.1) (Descending procedure). Let $R^S = \{j:j$ free and $c^*x^S + c^*_j+G_2(\alpha_b-\alpha_S-\alpha_j)<\hat{z}\}$. If $R^S \cap T^S$ is empty, S is fathomed and go to (2.2). If R^S contains all of the free variables, then augment S on the right by j (i.e. set $x_j = 1$) where x_j is a free variable which minimizes

$$(23) \qquad \sum_{i=1}^{m} \max \{-(y_i^S + a_{ij}, 0)\} + \sum_{\substack{i=1 \\ \text{and} \\ i\epsilon V_B}}^{m} \max \{(y_i^S + a_{ij} -1, 0\}.$$

In case of ties, choose any free variable with minimal c^*_j. If $\phi - R^S \sim S^c$ and $\delta > \delta_0$, augment S on the right by j for j which satisfies (23) breaking ties as before. If $\delta < \delta_0$, augment S on the right by $-j$ where j is that

free variable such that $c^*x^S + c_j^* + G_2(\alpha_b - \alpha_S - \alpha_j) > \hat{z}$ and (23) is a minimum. Ties may be broken arbitrarily. Set $R=R^S$. Go to (1.1) with the augmented partial solution and R.

(2.2) (Ascending procedure) Locate the rightmost element of S which is not underlined. If none exists, terminate. Otherwise, replace the element by its underlined complement and drop all elements to the right. If an overlined element is dropped, erase the stored solution of $G_2(\lambda_k)$, $k=0,1,\ldots,D$. Go to (1.1) with the new partial solution and $R=R^{\phi}$.

5. Numerical Example

The algorithm of section 4 is applied to the following problem which is presented in detached coefficient form in Table I. We want to minimize z, subject to $x_j=0$ or 1, $j=4,5,\ldots,14$, x_j non-negative integer, $j=1,2,3$.[1]

	x_1	x_2	x_3	x_4	x_5	x_6	x_7	x_8	x_9	x_{10}	x_{11}	x_{12}	x_{13}	x_{14}	z	b
	-1	0	0	1	1	2	-1	4	2	1	0	5	1	-2		7
(24)	0	0	-1	0	5	1	12	-1	-2	0	4	-3	1	8		2
	0	-1	0	1	1	3	3	4	1	2	1	4	0	1		6
				518	1200	1190	1996	1526	402	596	750	30	8	17	-48	0

TABLE I

The last three columns of A constitute an optimal LP basis. The basis is used to transform (24) into the form of (9). For convenience, all fractions are cleared by multiplying by D=48 and the zero-one requirement for y_1, y_2, y_3 becomes $y_i=0$ or 48.

The transformed problem is shown in Table II. For ease of exposition, the slacks are also made zero-one variables (the algorithm of Appendix D finds an initial feasible solution which actually rules out use of the slacks altogether).

basic variables	b	x_1	x_2	x_3	x_4	x_5	x_6	x_7	x_8	x_9	x_{10}	x_{11}
y_1	65	-1	-10	1	11	6	31	17	45	14	21	6
y_2	67	-35	34	-13	1	66	-19	19	-9	10	-33	18
y_3	28	4	-8	-4	4	24	20	76	12	-8	12	24
z	2962	242	164	142	112	84	72	62	44	38	26	18

TABLE II

[1] Strictly speaking, the coefficients c_j should be integers. They are fractions here solely for convenience and there is no loss of generality.

The group G is cyclic and has 48 elements. The relevant information needed for solving the various group problems is given in Table III. The righthand b is mapped into $\alpha_b = 17 \pmod{48}$

j	α_j	c^*_j
1	47	242
2	38	164
3	1	142
4	11	112
5	6	84
6	31	72
	17	62
8	45	44
9	14	38
10	21	26
11	6	18

Table III

The application of IP Algorithm I to the group problem yields $G_1(17)=62$. The backtracked solution with this value is $x_7=1$, $x_j=0$, $j\neq 7$. This solution is not feasible. The next step is to apply the backtracking algorithm of Appendix D to find an initial feasible solution. The solution it finds is $x_4=1$, $x_{11}=1$, $x_j=0$, $j\neq 4,11$, with a value of 130. The performance of the algorithm of section 4 is shown in Table IV.[1] An x in any column

[1]The parameter δ_0 was set equal to zero.

indicates that the corresponding bound or quantity is not needed. The optimal solution is

$$x_8=1,\ x_9=1,\ x_{11}=1,\ x_{13}=1,$$

$x_j=0$, otherwise, and the minimal cost is 64.

S	y^S	α_S	$\alpha_b-\alpha_S$	$G_1(\alpha_b-\alpha_S)$	$G_2(\alpha_b-\alpha_S)$	$G_3(\alpha_b-\alpha_S)$	$c{*}x^S$	z	R^S
ϕ	(65,67,78)	θ	17	62	62	62	0	130	{7,8,9,10,11}
{11}	(59,49,4)	6	11	82	82	82	18	130	{7,8,9,10}
{11,9}	(45,39,12)	20	45	44	44	44	56	130	{7,8,10}
{11,9,8}	(0,48,0)	x	x	x	x	x	100	100	x
$\{11,9,-8\}$	(45,39,12)	20	45	44	x	x	56	100	x
$\{11,-\underline{9},-\underline{8}\}$	(59,49,4)	6	11	82	x	x	18	100	x
$\{-\underline{11},-\underline{9},-8\}$	(65,67,28)	θ	17	62	62	62	0	100	ϕ

TABLE IV

6. Computational Results

In this section we report on some partial computational results with IP Algorithm I. We feel that these results offer some justification for the hypothesis that the main algorithm of this paper will prove computationally efficient. The encoding of the algorithm is currently under consideration.

The times listed in Table V are for the IBM 360-ASP system at M.I.T. The reader should note that the times of columns (a), (b), and (c) represent set-up times that are incurred only once for each integer programming problem. Columns (d) and (e) represent the variable times that would be incurred whenever IP Algorithm I is applied to problems of the form of (15).

A second point is that for problems 6 and 7, IP Algorithm I found an unconstrained shortest route path in the network with cost equal to the minimal cost of (9), but the backtracked solution was infeasible. This suggests that the algorithm should find or attempt to find all shortest route paths connecting θ to α_b. The necessary modifications to IP Algorithm I are discussed in [25].

TIMES ARE IN SECONDS*

Problem Number	Description	Ref.	Number of Rows	Number of Variables	D	Minimal LP Cost	$G_1(\alpha_b)$	Minimal Integer Cost	(a)	(b)	( )	(d)	(e)
1	Travelling Salesman	[5]	12	24	9	109.67	111.00	?	0.15	0.25	0.06	0.02	0.0
2	Travelling Salesman	[5]	31	55	15	56.20	61.00	63.00	1.91	3.39	0.36	0.04	0.03
3	Truck Routing	[22]	5	31	3	33.67	41.00	?	0.05	0.07	0.62	0.01	0.02
4	Test Problem	[27]	5	9	7	0.0	-7.00	-7.00	**	0.02	0.01	0.02	0.02
† 5	Fixed Charge 1	[14]	4	9	183	-8.79	-7.00	-7.00	0.01	0.01	0.0	0.37	0.0
6	Fixed Charge 2	[14]	4	9	258	-9.61	-8.00	-8.00	0.02	0.02	0.0	0.52	0.0
7	Fixed Charge 3	[14]	4	9	320	-11.81	-10.00	-10.00	0.02	0.01	0.01	0.77	0.0
8	Fixed Charge 4	[14]	4	9	205	-11.66	-11.00	-11.00	0.02	0.02	0.0	0.43	0.0
† 9	Fixed Charge 9	[14]	6	12	[illegible]00	-12.00	-9.00	-9.00	0.04	0.03	0.05	5.66	0.0
†10	IBM Test 1	[14]	7	14	32	7.50	8.00	8.00	0.05	0.05	0.06	0.07	0.0
11	IBM Test 2	[14]	7	14	32	5.75	7.00	7.00	0.05	0.05	0.0[illegible]	0.05	0.0
12	IBM Test 3	[14]	3	7	72	179.78	182.00	187.00	0.0	0.01	0.0[illegible]	0.13	0.0

* A listed time of 0.0 seconds indicates a time time of < 0.005

** This time is unavailable

† These problems were solved by IP Algorithm I

(a) Time for the simples algorithm, (b) time to convert to the group problem (c) time to transform group problem to canonical form, (d) time to find shortest route solution (e) time to backtrack shortest route path

TABLE V

7. Conclusion

It is hoped that the algorithm of this paper will prove computationally efficient for a wide variety of large zero-one integer programming problems. We feel that the computational experience gained thus far is promising, and the encoding of the remainder of the algorithm is underway. We emphasize, however, that an important next step in the development of algorithms such as the one here is the construction of a sequential decision process for controlling the bounding and branching procedures. On the basis of the characteristics possessed by a current partial solution, such a decision process would choose the bounds to be computed and those to be ignored. The decision process should also choose the next partial solution to be evaluated. The bounding and branching decisions would be made according to the optimality criterion of minimal expected cost to solution.

Thus, an additional consequence of the group theoretic approach are the qualitative insights it provides. For example, if the group is cyclic, and D is large and for most of the non-basic surplus variables the c_j^* are large, then the lower bounds from the partially constrained problem (18) will be high and solving this problem is worthwhile. Conversely, if D is small and c_j^* for most of the slack variables is small, then the computational effort required to solve (18) may not be justified. Many other insights come to mind but unfortunately, there is insufficient space here to discuss them.

Of course, most of the qualitative insights of [7] are relevant

here. As Geoffrion points out, a more flexible search within the framework of implicit enumeration is possible. He also mentions the possibility of using prior information to make a better start.

As discussed in [25], D may be quite large for some problems, and this can cause computational difficulties. If so, a different algorithm for solving (4) which concentrates on calculating only $G_1(\alpha_b)$ (or $G_2(\alpha_b - \alpha_s)$) may be suggested. A second approach may be to try to decompose the integer programming problem, or the group network, into more manageable pieces. This is an area for future research. A third alternative is the following. Without loss of generality, assume (1) has the property that $c_j' \geq 0$. This condition implies that the initial solution $x'=0, s=b$ to (2) is dual feasible, and D for the corresponding basis (I) is one. Beginning with this dual feasible solution, iterate with the dual simplex algorithm until (2) is solved or the basis determinant becomes large, but not too large, say $300 < D < 3000$. Use the dual feasible basis with this determinant value to transform (2) into (4) and (9). The resulting c_j^* and $c_{j_k}^*$ are non-negative and hence there is no difficulty with the shortest route and implicit enumeration formulations.

Another area for future research is the synthesis of group theory and special purpose branch and bound algorithms ([15], [18], [19], [22]).

REFERENCES

[1] Balas, E., "An Additive Algorithm for Solving Linear Programs with Zero-one Variables," Operations Research, 13, pp. 517-546, (1965).

[2] Balas, E., "Discrete Programming by the Filter Method," Operations Research, 15, pp. 915-957, (1967).

[3] Balinski, M.L., "Integer Programming: Methods, Uses, Computation," Management Science, 12, pp. 253-313, (1965).

[4] Benders, J.F., "Partitioning Procedures for Solving Mixed-Variable Programming Problems," Numerische Mathematik, 4, pp. 238-252, (1962).

[5] Datnzig, G.B., Linear Programming and Extensions, Princeton University Press, 1963.

[6] Fuchs, L., Abelian Groups, Pergamon Press, 1960.

[7] Geoffrion, A.M., "Integer Programming by Implicit Enumeration and Balas' Method," SIAM Review, 9, No. 2, pp. 178-190, (1967).

[8] Geoffrion, A.M., "Implicit Enumeration Using an Imbedded Linear Program," the RAND Corporation, RM-5406-PR, September, 1967

[9] Geoffrion, A.M., "Recent Computational Experience with Three Classes of Integer Linear Programs," The RAND Corporation, P-3699, October, 1967.

[10] Glover, F., "An Algorithm for Solving the Linear Integer Programming Problem Over a Finite Additive Group, with Extensions to Solving General Linear and Certain Nonlinear Integer Problems," Operations Research Center Report 66-29, University of California, Berkeley (1966).

[11] Gomory, R.E., "An Algorithm for Integer Solutions to Linear Programs," Recent Advances in Mathematical Programming, McGraw-Hill Book Company, New York, 1963, p. 269.

[12] Gomory, R.E., "On the Relation Between Integer and Non-Integer Solutions to Linear Programs," Proceedings of the National Academy of Sciences, Vol. 53, pp. 250-265 (1965).

[13] Gomory, R.E., "Faces of an Integer Polyhedron", Proceedings of the National Academy of Sciences, Vol. 57, pp. 16-18, (1967).

[14] Haldi, J., "25 Integer Programming Test Problems," Working Paper No. 43, Graduate School of Business, Stanford University, 1964.

[15] Ignall, E. and Schrage, L, "Application of the Branch-and-Bound Technique to Some Flow-Shop Scheduling Problem," Operations Research, 11, pp. 972-989, (1963).

[16] Lawler, E.L. and D. E. Ward, "Branch-and-Bound Methods: A Survey", Operations Research, 14, pp. 699-719, (1966).

[17] Lembe, C.E. and K. Spielberg, "Direct Search Algorithms for Zero-One and Mixed-Integer Programming," Operations Research, 15, pp. 892-914, (1967).

[18] Little, J.D.C., Murty, K.G., Sweeney, D.W., and Karel, C., "An Algorithm for the Travelling Salesman Problem," Operations Research, 11,pp. 972-989, (1963).

[19] Lomnicki, Z.A., "A Branch and Bound Algorithm for the Exact Solution of the Three-Machine Scheduling Problem," Operational Research Quarterly, 16, pp. 89-100, (1965).

[20] Kolesar, P.J., "A Branch and Bound Algorithm for the Knapsack Problem," Management Science, 13, pp. 723-735, (1967).

[21] Mostow, G.D., J.D. Sampson and J.P. Meyer, Fundamental Structures of Algebra, McGraw-Hill, 1963.

[22] Pierce, J.F., "Application of Combinatorial Programming to a Class of All-Zero-One Integer Programming Problems," IBM Cambridge Scientific Center Report, Cambridge, Massachusetts, July, 1966.

[23] Shapiro, J.F. and H.M. Wagner, "A Finite Renewal Algorithm for the Knapsack and Turnpike Models," Operations Research, 15, No. 2, pp. 319-341, (1967).

[24] Shapiro, J.F., "Finite Renewal Algorithms for Discrete Dynamic Programming Models," Technical Report No. 20, Graduate School of Business, Stanford University, (1967).

[25] Shapiro, J.F., "Dynamic Programming Algorithms for the Integer Programming Problem - I: The Integer Programming Problem Viewed as a Knapsack Type Problem," to appear in Operations Research.

[26] Shapiro, J.F., "Dynamic Programming Algorithms for the Integer Programming Problem - II: Extension to a General Integer Programming Algorithm", submitted to Operations Research.

[27] White, W.W., "On a Group Theoretic Approach to Linear Integer Programming, Operations Research Center Report 66-27, University of California, Berkeley, (1966).

APPENDIX A - IP ALGORITHM I

This appendix contains a statement of the algorithm from [25] for solving the group problem (3). Strictly speaking, the algorithm below is a slight modification of the original. The differences are relatively slight, however, and the reader is referred to [25] for an intuitive explanation. The only remark we make here is that node λ_D has been added to the network representation. The algorithm finds the shortest route path from θ to λ_D which is the least cost elementary circuit in the network.

Shortest Route Algorithm for the Unconstrained Group Problem

STEP 1 (Initialization) Set $Z(\lambda_k) = Z(D+1)$ for $k=1,2,\ldots,D$ where $Z = \max_{j=1,\ldots,n} c_j^*$; set $Z(0) = 0$. Also, set $j(\lambda_k) = 0$ for $k=0,1,\ldots,D$. For $j=1,\ldots,n$, and $\alpha_j \neq \theta$, set $Z(\alpha_j) = c_j^*$, $j(\alpha_j) = j$ and $a_{\alpha_j} = 2$ only if $c_j^* < Z(\alpha_j)$. For $j=1,\ldots,n$, and $\alpha_j = \theta$, set $Z(\lambda_D) = c_j^*$, $j(\lambda_D) = j$ only if $c_j^* < Z(\lambda_D)$. For all nodes λ_k for which a_{λ_k} is not specified set $a_{\lambda_k} = 1$. Go to Step 2 with $\lambda_k = \theta$.

STEP 2 Stop if $a_{\lambda_k} = 1$ for $k=0,1,\ldots,D$. Then $G_1(\lambda_k) = Z(\lambda_k)$ for $k=0,1,\ldots,D$. Otherwise (1) if there is a $k' > k$ such that $a_{\lambda_{k'}} = 2$, index k to k' or (2) if there is no $k' > k$ such that $a_{\lambda_{k'}} = 2$ index k to the smallest $k''(<k)$ with $a_{\lambda_{k''}} = 2$. Go to Step 3.

STEP 3 For $j=j(\lambda_k), j(\lambda_k)+1,\ldots,n$, and $\lambda_k+\alpha_j \neq \theta$, set $Z(\lambda_k + \alpha_j) = c_j^* + Z(\lambda_k)$,

$j(\lambda_k + \alpha_j) = j$ and $a_{\lambda_k+\alpha_j}$ 2 only if $c_j^* + Z(\lambda_k$ < $Z(\lambda_k + \alpha_j$ For $j=j(\lambda_k$, $j(\lambda_k$ + 1,...,n, and $\lambda_k + \alpha_j$ θ, set $Z(\lambda_D$ $c_j^* + Z(\lambda_j$ and $j(\lambda_D) = j$ only if $c_j^* + Z(\lambda_j) < Z(\lambda_D)$ Return to Step 2.

APPENDIX B - A BACKTRACKING ALGORITHM FOR THE ZERO-ONE GROUP PROBLEM

Our concern here is the construction of an algorithm for solving the group problem (3) with the side constraints $x_j=0$ or 1, $j \epsilon V_R$; namely,

$$\min \sum_{j=1}^{n} c_j^* x_j$$

subject to

$$\sum_{j=1}^{n} \alpha_j x_j = \alpha_b \pmod D$$

(25) $x_j=0$ or 1, $j \epsilon V_R$;

x_j non-negative integer, $j \epsilon U_R$.

In terms of the lower bounds used in the body of the paper, the algorithm here finds $G_3(\alpha_b)$. The algorithm can also be used to find $G_3(\alpha_b - \alpha_S)$ for some partial solution S by limiting the variables used in (25) to the free variables relative to S, and by changing the right hand side group element to $\alpha_b - \alpha_S$.

An intuitive explanation of the algorithm is the following. First, attempt to solve (25) by using IP Algorithm I which solves (4) for all right hand sides λ_k, $k=0,1,\ldots,D$. Backtrack each of the optimal paths from θ to λ_k. Some of the paths and the corresponding solutions to (3) are the correct form for (25) with right hand side λ_k. In other words, there must be some nodes λ_k for which we have a fortiori $G_1(\lambda_k) = G_2(\lambda_k)$. Let K_1 be the set of indices for which $G_1(\lambda_k) = G_3(\lambda_k)$ and assume $G1(\alpha_b) \neq G_3(\alpha_b)$.

The next step is to apply the variant of IP Algorithm I described in Appendix C to problem (25). This algorithm is intended to provide a good feasible solution to (25) and thereby reduce the work of the exact algorithm below. Let $\tilde{G}_3(\alpha_b)$ be the value of the feasible solution found by the heuristic algorithm and let this be the initial value of the incumbent $\hat{z}_3(\alpha_b)$. The reader is referred to Appendix C for a detailed description of that algorithm.

With the above background, the exact algorithm for solving (25) can be intuitively described. All circuitless paths[1] between θ and α_b can be found by a backtracking dynamic programming algorithm. An equivalent approach is to extend paths forward from θ to α_b. The first step is to extend the paths $\mu_1, \mu_2, \ldots, \mu_n$ from θ to α_j with costs c_j^* , $j=1,\ldots,n$. At any given time during the running of the algorithm, there will be a collection $\{\mu_t\}_{t=1}^{T}$ of paths being tested and extended.

Consider an arbitrary path μ_t beginning at θ whose last arc is $(\lambda_k - \alpha_{j_t}, \lambda_k)$. If $\lambda_k = \alpha_b$, then μ_t is terminated and its cost is compared with the cost of the incumbent path and if possible an improvement is made. If $\lambda_k \neq \alpha_b$, we use the form of an Optimal Path lemma[2] to extend μ_t. Specifically, μ_t is extended by the arcs $(\lambda_k, \lambda_k + \alpha_j)$ either for $j \leq j_t - 1$ if $j_t \varepsilon V_R$, or for $j \leq j_t$ if $j \varepsilon U_R$. Thus it appears that the path μ_t generates $j_t - 1$ or j_t new paths.

---

[1]Since all arc costs $c_j^* \geq 0$, we can without loss of optimality limit the search to circuitless paths.

[2]See [25].

Fortunately, it may not be necessary to extend μ_t by all the arcs indicated above and in some cases, it may be possible to discontinue μ_t First, we draw the arc $(\lambda_k, \lambda_k + \alpha_j)$ only if $c(\mu_t) + c_j^* < \hat{z}_3(\alpha_b)$ where $\hat{z}_3(\alpha_b)$ is the value of the incumbent.

There are three additional tests which can be used to terminate μ_t. First, if the paths λ_{t1} and λ_{t2} are drawn to λ_k with last arcs $(\lambda_k - \alpha_{j_{t1}}, \lambda_k)$ and $(\lambda_k - \alpha_{j_{t2}}, \lambda_k)$ respectively, and $c(\mu_{t1}) < c(\mu_{t2})$ while $j_{t1} \geq j_{t2}$, then μ_{t2} can be discontinued. This is because μ_{t1} has more free variables to use in reaching α_b than μ_{t1} and it is already at least as good.

Among the paths to λ_k which remain, if there is a μ_t such that

$$c(\mu_t) + G_1(\alpha_b - \lambda_k) \geq \hat{z}_3(\alpha_b),$$

then μ_t need not be extended since $G1(\alpha_b - \lambda_k)$ is a lower bound on the cost of extending μ_t from λ_k to α_b. Conversely, if

$$c(\mu_t) + G_1(\alpha_b - \lambda_k) < \hat{z}_3(\alpha_b),$$

and $G_1(\alpha_b - \lambda_k) = G_3(\alpha_b - \lambda_k)$, and $x^1(\alpha_b - \lambda_k)$ uses only free variables relative to μ_t, then $z_3(\alpha_b) \leftarrow c(\mu_t) + G_1(\alpha_b - \lambda_k)$. The solution with value $\hat{z}_3(\alpha_b)$ is found by backtracking μ_t and adding the resulting solution to $x^1(\alpha_b - \lambda_k)$. Once again the path μ_t need not be extended.

We remark that an explicit numerical label for μ_t is superfluous. Instead, without loss of generality, we need record only λ_k, j_t, and

$v_t = c(\mu_t)$. To reconstruct μ_t from this information, we look through the paths drawn to $\lambda_k - \alpha_{j_t} = \lambda_\ell$ for one with value $v_s = v_t - c_j^*$ and last arc $j_s \geq j_t + 1$ if $j_t \varepsilon V_R$, or $j_s \geq j_t$ if $j_t \varepsilon U_R$.

The algorithm below can easily be ammended to find $G_3(\lambda_k)$, k=0,1,...,D. This variant of the algorithm may often be preferred, especially at the top of the tree of enumerated solutions ($S=\phi$).

<u>Backtracking Algorithm for the Group Problem with Zero-one Constraints</u>

STEP 1 (Initialization) Set $\hat{z}_3(\alpha_b)$ equal to the value of the best path from θ to α_b found by the heuristic algorithm. Let $\hat{z}_3(\alpha_b)$ be the corresponding solution. Put the paths $(\alpha_j;j;v_j;2)$ on the node lists α_j, j=1,...,n, and set $a_j=2$. Let $a_k=1$ for all other nodes λ_k. Go to step 2 with k=0.

STEP 2 If $a_k=1$, k=0,1,2,...,D-1, then the optimal solution is the incumbent solution $\hat{x}^3(\alpha_b)$ with cost $z_3(\alpha_b)$. If there is an $a_k=2$, try to index k to the lexicographically smallest k' > k such that $a_{k'}=2$; otherwise, index k to the lexicographically smallest k' such that $a_{k'}=2$. Go to Step 3.

STEP 3 Set $a_k=1$. Let

$$J_K = \{j \mid (\lambda_k;j;v_j;\delta) \text{ such that } \delta = 2$$

In particular, suppose

$$J_K = \{j_1,j_2,\ldots,j_s \mid j_1 \leq j_2 \leq \cdot \cdot < j_s \text{ and } v_{j_i} \geq v_{j_{i+1}} \text{ if } j_i = j_{i+1} \} .$$

Reduce J_k by the following sequence of tests.

Test 1: For i=1,...,s, if $v_{j_i} \geq \hat{z}_3(\alpha_b)$, remove j_i from J_k and erase the path $(\lambda_k;j_i;v_{j_i};2)$. If $J_k = \phi$, return to Step 2. Otherwise, go to test 2 (There are s_1 paths remaining in J_k).

Test 2: For i=1,..., s_1, if $v_{j_i} \geq v_j$, ℓ=i+1,...,s_1, remove j_i from J_k and erase the path $(\lambda_k,j_i,v_{j_i},2)$. If $J_k = \phi$ return to Step 2. Otherwise, go to test 3 (there are s_2 paths remaining in J_k).

Test 3: For i=1,...,s_2, if $v_{j_i} + {}_{G1}(\alpha_b-\lambda_k) \geq \hat{z}_3(\alpha_b)$, remove j_i from J_k and erase the path $(\lambda_k,j_i,v_{j_i},2)$. If

(i) $v_{j_i} + {}_{G1}(\alpha_b-\alpha_k) \leq \hat{z}_3(\alpha_b)$,

and

(ii) ${}_{G1}(\alpha_b-\lambda_k) = G_3(\alpha_b-\lambda_k)$,

and

(iii) either $j(\alpha_b-\lambda_k) \leq j_i-1$ if $j_i \varepsilon V_R$,

of $j(\alpha_b-\lambda_k) \leq j_i$ if $j_i \varepsilon U_R$.

Then $\hat{z}_3(\alpha_b) \leftarrow v_{j_i} + {}_{G1}(\alpha_b-\lambda_k)$ and the new incumbent is found by backtracking the path $(\lambda_k;j_i;v_{j_i};2)$ and adding this solution to $x^1(\alpha_b-\alpha_k)$. Remove j_i from J_k and erase the path $(\lambda_k,j_i,v_{j_i},2)$.

If $J_k=\phi$, return to step 1. Otherwise, go to step 4 with s_3 paths remaining in J_k.

STEP 4 For i=1,2,...,s_3, set δ_i=1. Extend the path $(\lambda_k,j_i,v_j,1)$ to the

paths $(\lambda_k+\alpha_h, h, v_{ji} + c_h, 2)$ for either $h=1,\ldots,j_i-1$ if $j_i \varepsilon V_R$, or for $h=1,\ldots,j_i$ if $j_i \varepsilon U_R$. For each node $\lambda_k+\alpha_h = \lambda_\ell$ thereby spanned, set $a_\ell = 2$ if necessary. Return to Step 2.

APPENDIX C - HEURISTIC ALGORITHM FOR THE ZERO-ONE GROUP PROBLEM

It is easily seen that the algorithm of Appendix A can be ammended to find feasible solutions to problem (19). This is accomplished by changing Step 3 so that the arc $(\lambda_k, \lambda_k + \alpha_{j(\lambda_k)})$ is not drawn when $j(\lambda_k) \varepsilon V_R$.

In order that this algorithm have maximum effectiveness, reorder temporarily the non-basic variables so that

(1) The non-basic surplus variables have the highest index. Within this set, order by decreasing value of c_j^*.

(2) The non-basic zero-one variables are indexed bv decreasing value of c_j^* except when several variables are mapped into the same group element. In the latter case, all of the variables mapped into the same group element should be indexed consecutively by <u>increasing</u> value of c_j^*.

APPENDIX D - BACKTRACKING ALGORITHM

The algorithm below generates paths from θ to α_b and tests the corresponding solutions (corrections) for feasibility in (9). Its main purpose is to find a good starting solution for the algorithm in section 4, rather than an optimal solution. Since short paths tend to have lower cost than long paths, it is felt that this algorithm will generate good feasible solutions. It is possible, however, for the algorithm to discover and identify an optimal solution.

As previously mentioned, it may be preferable for some problems to use the algorithm of [7] to find an initial feasible solution. The algorithm here will be particularly effective, of course, for problems which have few non-basic variables, say $n < 10$.

Two comments about the algorithm are in order. The parameter β is chosen by the user to prevent excessive backtracking. Roughly speaking, it has been estimated that an analysis involving 10,000 triples would require 100,000 bytes. Thus, the 360-ASP system could handle up to 25,000 triples. Second, the algorithm can be limited to finding only an initial feasible solution by exiting from Step 3 whenever $z(\alpha_b) < +\infty$.

STEP 1 (Initialization): Set K=1 and $\hat{z}(\alpha_b) = +\infty$, $\hat{x}(\alpha_b) = (0,0,\ldots,0)$ Add the triples (j,α_j,c_j^*) to the K-list. Go to Step 2.

STEP 2 If the K-list is empty, then $\hat{x}(\alpha_b)$ with value $\hat{z}(\alpha_b)$ is an optimal correction in (9). If the K-list is not empty, perform the following analysis for each triple (j,λ_k,v). If $\lambda_k = \alpha_b$, go to Step 3 with the triple

(j,λ_k,v). Otherwise, either for each $\ell \leq j$ if $j \epsilon U_R$, or for each $\ell \leq j-1$ if $j \epsilon V_R$, place $(\ell, \lambda_k+\alpha_\ell, v+c_\ell^*)$ on the (K+1) list only if $K < \beta-1$ and $v+c_\ell^* < \hat{z}(\alpha_b)$.

After all triples on the K-list are analyzed, index K to K+1 and repeat only if $K < \beta-1$. If $K=\beta$, terminate with $\hat{z}(\alpha_b)$ and $\hat{x}(\alpha_b)$.

STEP 3 Search through the triples on the (K-1) list for a triple $(\ell$ $\lambda_k-\alpha_j, v-c_j^*)$ such that either $\ell \geq j$ if $j \epsilon U_R$, or $\ell \geq j+1$ if $j \epsilon V_R$. Search through the triples on the (K-2) list for a triple $(w, \ell_k-\alpha_j-\alpha_\ell, v-c_j^*-c_\ell^*)$ such that either $w>\ell$ if $\ell \epsilon U_R$, or $w > \ell+1$ if $\ell \epsilon V_R$. Continue backtracking until the entire path is found. Test the correction $x=(x_1,\ldots,x_n)$ for feasibility in (9) where x_j = number of times α_j is used in the backtracked path. If x is feasible, and $\sum_{j=1}^{n} c_j^* x_j < \hat{z}(\alpha_b)$, $\hat{z}(\alpha_b) \leftarrow \sum_{j=1}^{n} c_j^* x_j$ and $\hat{x}(\alpha_b) \leftarrow x$. Return to Step 2.

CPSIA information can be obtained at www.ICGtesting.com
Printed in the USA
LVOW10s1317100816

499828LV00031B/627/P

9 781332 262533